WATSON

My Garden Grows

TO

My Pretty Maids

all in a row

NANCY, WENDY, CLYDE, LINDA, ANN, NANCY CAMERON, *and* CAITLIN

and to NORMAN M. DINGMAN, *Grandfather of them all*

MY GARDEN GROWS

WRITTEN *and* ILLUSTRATED BY

Aldren A. Watson

NEW YORK THE VIKING PRESS

First published in 1962 by The Viking Press, Inc.
625 Madison Avenue, New York 22, N.Y.
Published simultaneously in Canada by The Macmillan Company of Canada Limited

Third printing January 1967

Library of Congress catalog card number: 61-11674

635 1. Plants
2. Vegetable gardening

PRINTED IN THE U.S.A.

It is spring in my garden. The birds sing spring in the trees, and the sun shines spring-warm and bright. I dig into the brown, crumbly earth with my spade and watch the earthworms wriggle under my bare feet. I turn the fresh earth upside down and watch the beetles tumble out.

There is a lot to do. Before the seeds can be planted, the earth has to be raked and raked until all the big lumps are gone and it is nice and fine. Then the threadlike roots which come from the seeds can easily push through the soft ground. The trouble is, weeds like the soft, crumbly earth too. The weeds must be pulled out quite often. Otherwise they will grow faster than the vegetables, and there will be nothing but weeds in the garden.

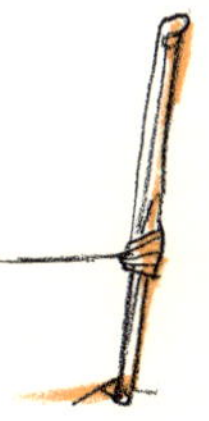

Now, with the handle of a hoe, I make straight rows, guided by a string stretched between two stakes. Plop, plop, plop, into the row I drop the seeds. After I plant the seeds I will rake the crumbs of dirt into the row on top of the seeds and tamp the row down gently, so the dirt presses against the seeds on every side. This will make them start growing faster.

The beet seeds are funny, crinkly things. They look like some kind of breakfast cereal. Although they are quite big compared to some other seeds, they are so light that a good wind will blow them out of my hand. Beets are "root" vegetables. This means that the part you eat grows under ground and stays there till you want to pick it. The beet is really the root of the plant. Carrots and potatoes are root vegetables too. When the beets are ready to pull, you can see just a little of the round beet poking above the ground. Above that are the green tops, which are also good to eat.

In my garden are carrots. The fine seeds are even lighter than beet seeds, and much smaller, like little new moons. The best-tasting carrots are those that have grown fast, and since they grow the fastest early in the spring, you should plant them as soon as the ground is ready. When carrots come up they look almost like grass. But as soon as they put out little branches they begin

to look like vegetables. The tops grow and grow, adding branches till they look like feather dusters. But where are the carrots? Under the ground, remember? Carrots are *root* vegetables. If you could crawl under the dirt and watch, you would see the bright orange carrot getting bigger and bigger every day. It grows down, longer and longer, until it's finished and you can pull it up.

In my garden is a toad. Whenever I plant seeds, or pull out weeds in my garden, I see him sitting in the shade, waiting to catch some bugs. He doesn't know it, but he is helping me. He spends all day snatching beetles and other insects with his wide mouth. It's a good thing he does, or the bugs would eat up all my vegetables. The toad must like me too, because he never hops away. He must think he lives in a forest.

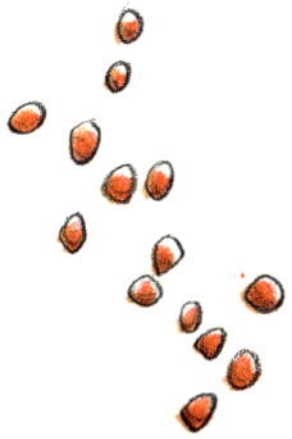

Right now he is sitting under a broccoli plant. Broccoli seeds are like tiny brown beads. A tablespoonful of them would make enough broccoli to fill up a whole garden. You'd never think the tiny seeds could grow into such big, tall plants. The little head is the part you can eat. When it gets fat and tight, you cut it off with a knife and cook it. The plant keeps on making more and more bumpy little broccoli heads, even though you cut them off.

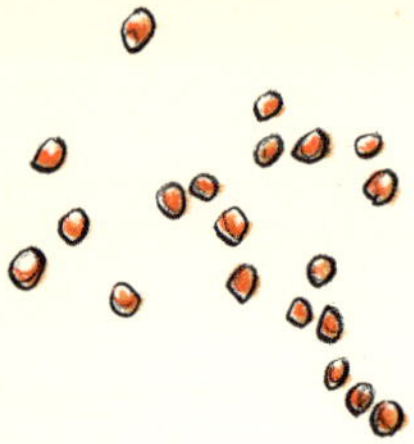

Cabbage grows in my garden. Like broccoli seeds, cabbage seeds are tiny dark brown beads. The plants are so much alike, it is very hard to tell them apart until they have grown pretty big and have several leaves. The broccoli leaves are a bluish green, instead of the grassier green of the cabbage. It takes a long time for a head of cabbage to get round and hard, the way it looks in the store. At first the leaves are loose. As more leaves crowd into the center, the head gets bigger, tighter, and more firm. At last it is ready to eat. Toad loves cabbage more than anything else on a hot August afternoon—not to eat, for I have never seen him eat any, but the big juicy, cool leaves make a nice tent to shade his back from the hot sun.

In my garden are cucumbers. Their seeds are little flat, pointed wafers. From looking at this seed you would have no way of knowing that it grows in such a crazy way, all over the garden. There are a lot of different kinds of cucumbers: big ones, little ones, straight and crooked ones. Some have sharp little spines on them. Pickles are *not* in my garden, but pickles are just plain cucumbers that Mother has soaked in a special salt water or vinegar juice.

In my garden is lettuce. It is one of the first things from the garden that can be used. It likes cool weather and a little rain every so often. Lettuce doesn't grow well in the middle of summer, because the weather is too hot. Lettuce is the start of almost every salad; in fact it makes a very good one all by itself.

It is surprising how many onions Mother uses, for soup, for salad, with meat, and in stew. I guess that's why we grow so many onions every year. Onion seeds are very small, coal-black chips. Where are the seeds? They are all in a little round spire at the very top end of the stalk. The plant begins very small, but grows fast in the cool spring ground. The shiny, waxy green tops make the garden look very tidy, growing in even rows, standing up so straight. They make me hungry for a salad, or a juicy hamburger with a slice of onion on top. My sister likes to pull the onions in August, and after they have dried she braids them into bunches. Then she hangs them in the shed to keep. The toad never seems to hop over by the onions, and I'm sure it must be because not one single bug ever bothers them.

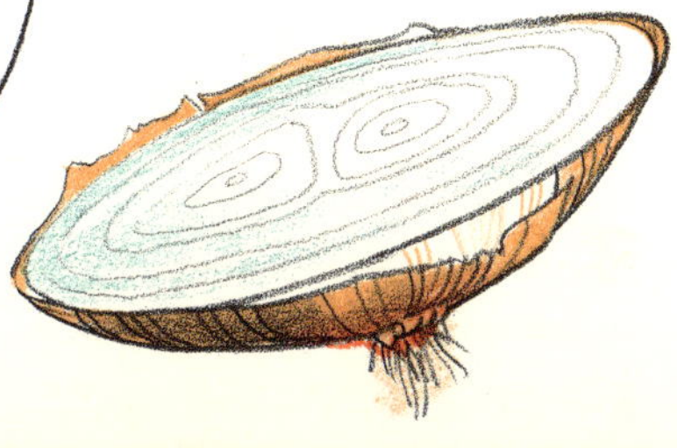

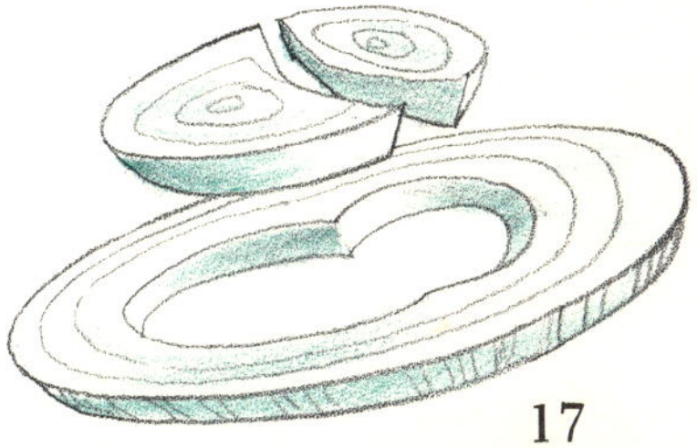

Peas are in my garden, too. A pea seed is the same size as the pea you eat. In fact, the peas *are* the seeds. They are all wrinkled because they have been dried for keeping. The pea vines grow about twice as tall as Nancy, so Daddy and Peter put wire on stakes to hold up the thick, climbing branches. They, too, are fast growers in the cool spring. Just as soon as the blossoms fall off, you can see the pods starting. They get longer and fatter. Inside, the peas are getting round and plump too. And when the pod is opened, there are all the peas in a row!

Corn is in my garden. The family likes corn so well that Mother and Daddy always plant more than we can eat. But when it's ready there are lots of neighbors to help us eat it. If you think broccoli makes a lot of plant from a tiny seed, turn the page on end and see what *this* little seed has done. Every corn stalk will

have one or two ears. Every ear has about three hundred kernels —the golden, tasty part you eat. These kernels are really the seeds. And each seed, if it were planted the next year, would make a separate stalk with one or two more ears on *it*. What do you think of that? At the tip end of the cornstalk roots you can see the

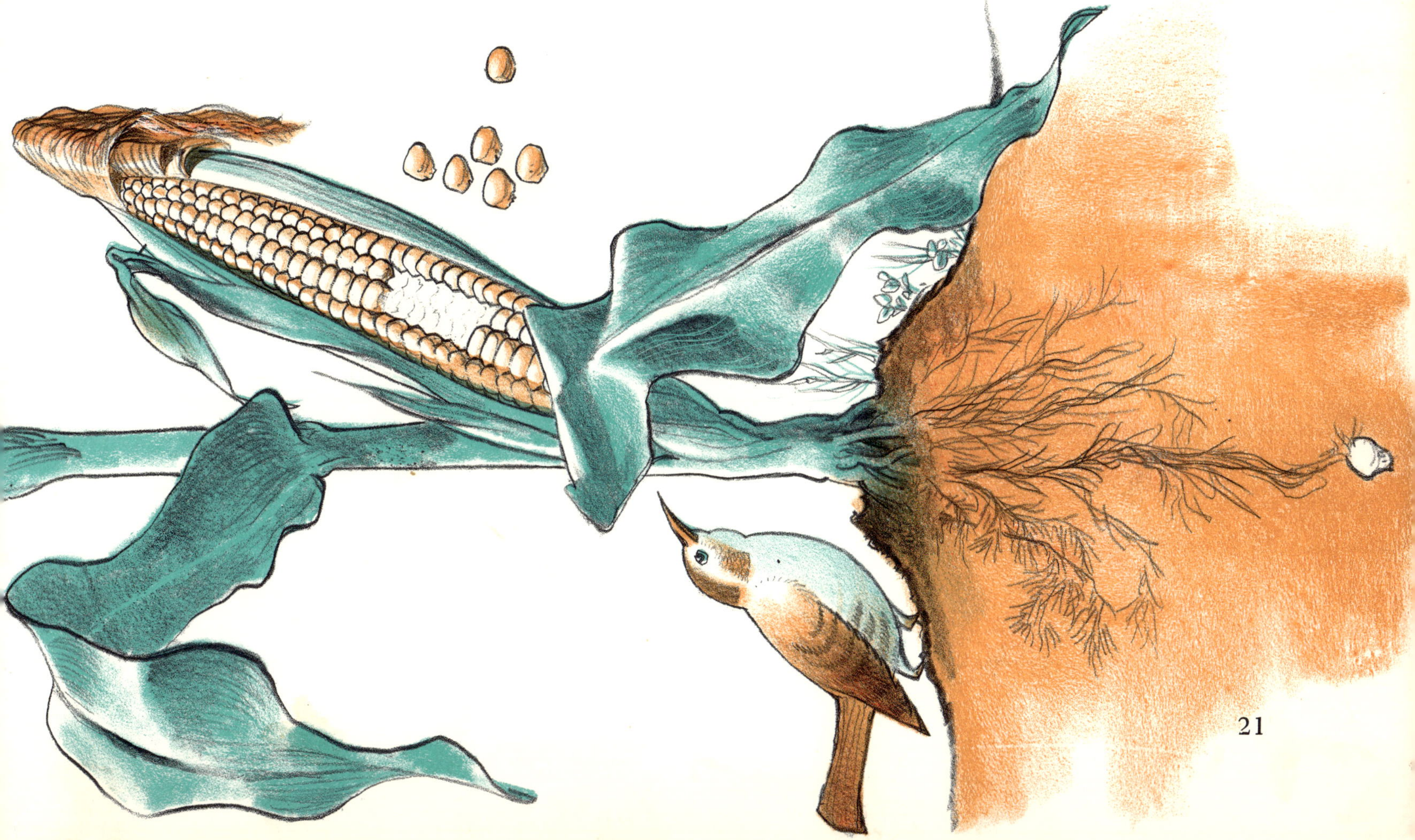

single kernel from which it grew. The Indians grew corn long ago, but the ears were smaller, and the seeds were of different colors. If you want to have some fun, plant a packet of Indian corn. In the fall, when it is ripe, tie several ears together into a bunch and hang it up for a Halloween or Thanksgiving decoration.

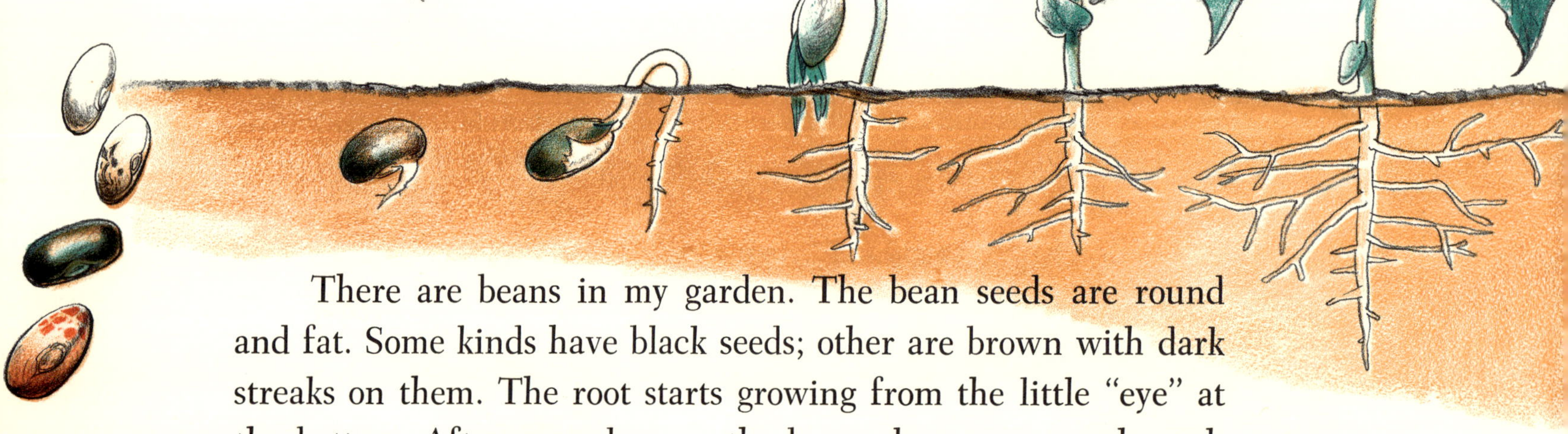

There are beans in my garden. The bean seeds are round and fat. Some kinds have black seeds; other are brown with dark streaks on them. The root starts growing from the little "eye" at the bottom. After a week or so the bean plant pops up through the ground. But there aren't any beans on it, not yet. More leaves grow, till at last it is a bush large enough to blossom. Each blossom will become a bean pod. When it is time to pick the beans for supper, they will hang down in thick clusters on the bush. Most beans we eat are not really finished. The round beans inside the shell have not got as big as they would if left on the bush. But when eaten this way, they are much more tender. If you want to use just the inside beans, you can leave them on the bush till fall, and then shell out the beans, throwing away the pod. These are called "shell beans."

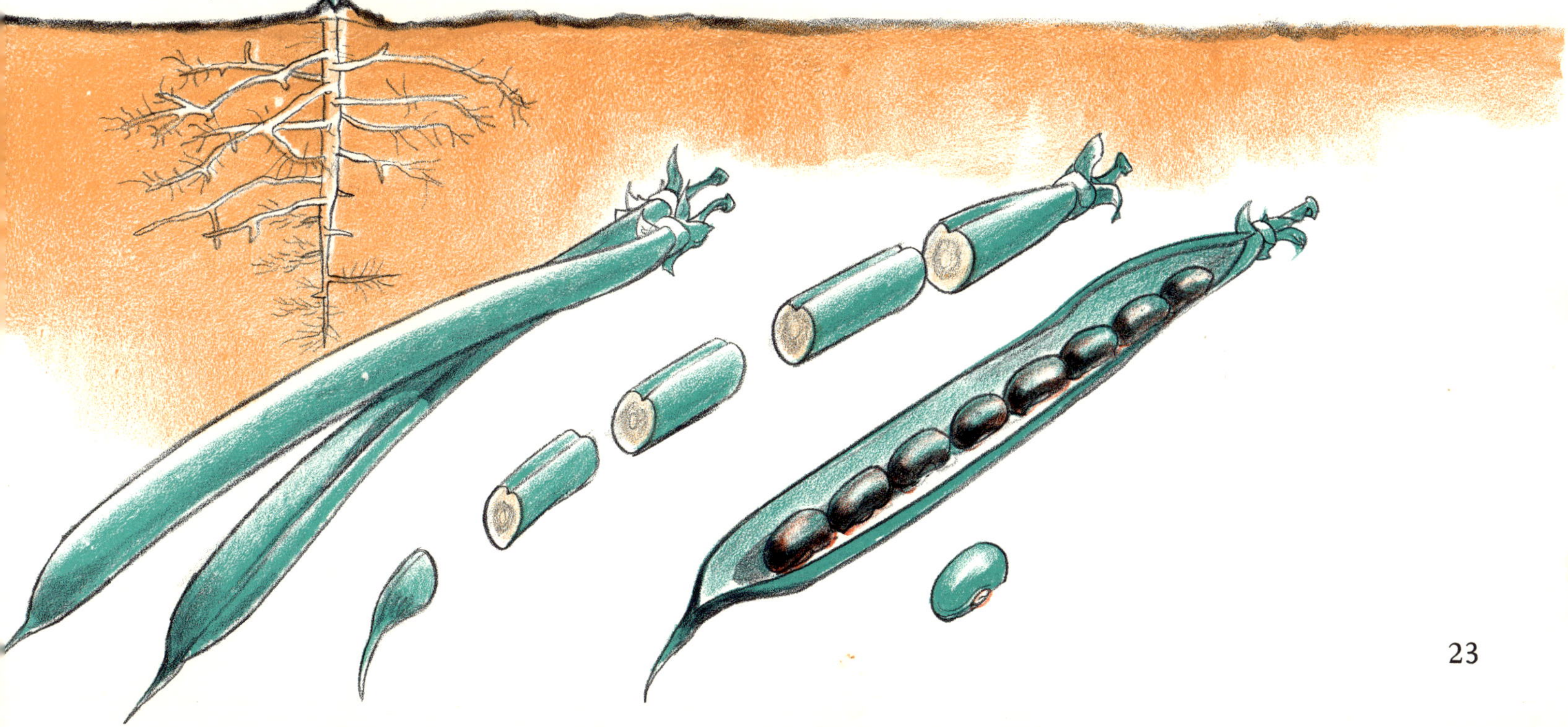

This eggplant is in my garden. Does it look like an egg? I don't know why it is called *egg*plant, but it's very, very good to eat. The baby sits down and holds a big eggplant in her lap. It's just about as big as her head. Although eggplants are quite heavy, they hang up under the fuzzy pointed leaves of the plant. Sometimes a few rest on the ground. They look like giant purple plums. If only plums would grow that big!

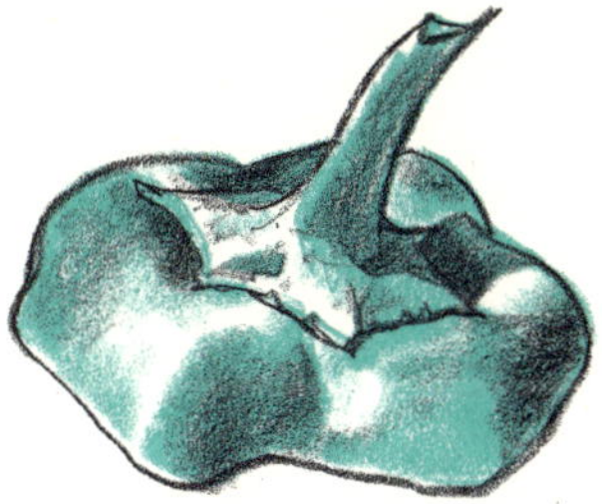

In my garden are peppers. They are just the opposite of peas: instead of eating the seeds and throwing away the pod, you eat the pod and throw away the seeds. Each plant has five or six peppers, hanging on the under side of the little bush. These are green peppers, but there is also a red kind. All of them are as shiny as though they had been waxed. Did you ever eat stuffed peppers?

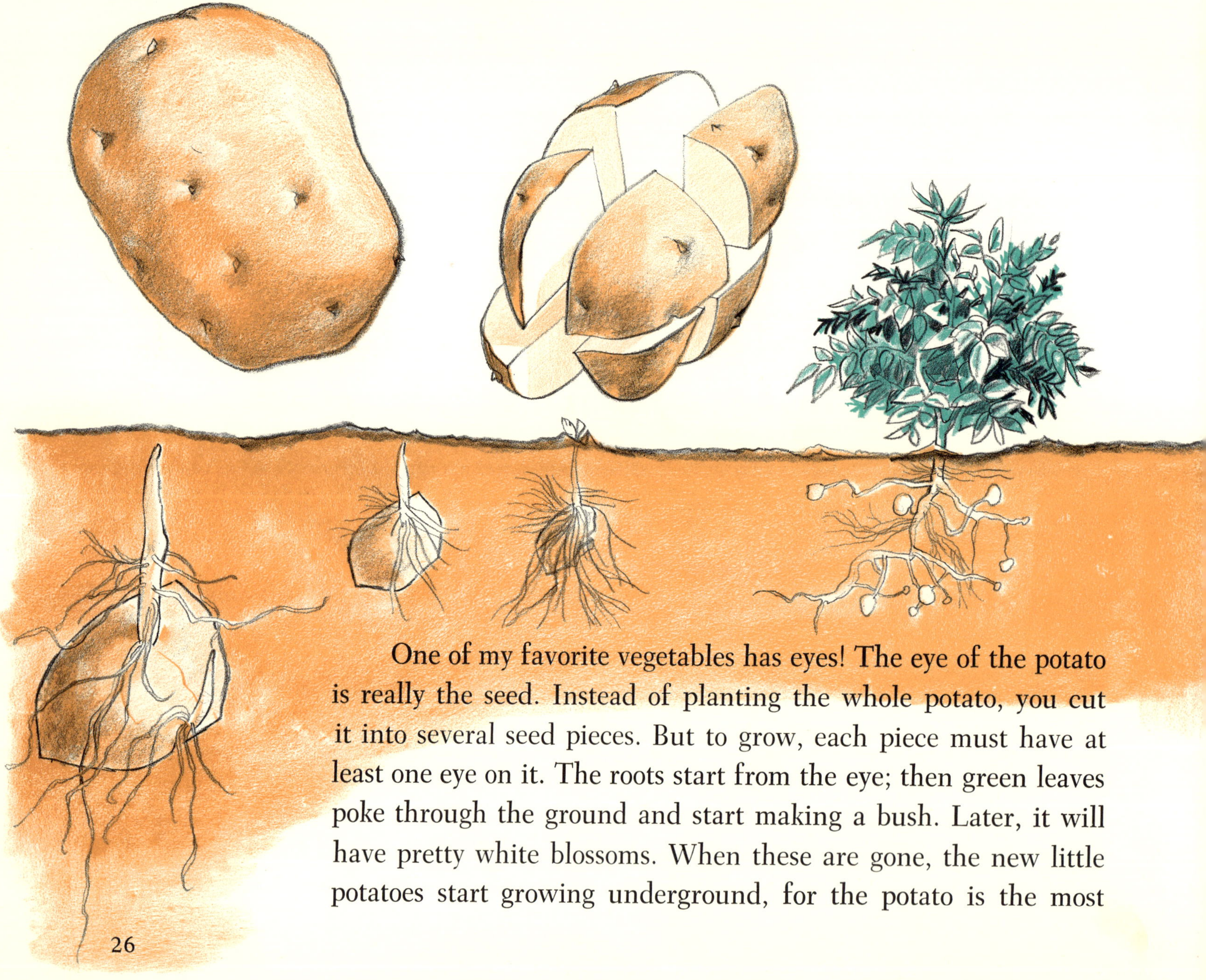

One of my favorite vegetables has eyes! The eye of the potato is really the seed. Instead of planting the whole potato, you cut it into several seed pieces. But to grow, each piece must have at least one eye on it. The roots start from the eye; then green leaves poke through the ground and start making a bush. Later, it will have pretty white blossoms. When these are gone, the new little potatoes start growing underground, for the potato is the most

famous root vegetable. At the end of summer the tops dry up and flop over on the ground. Now you can tell that the potatoes are ready to dig. Each plant will form about five or six nice fat brown potatoes in a cluster on the roots. When we dig ours, Caitlin likes to sit in the dirt and collect the little ones that didn't have time to get finished. They make good marbles for her to play with. And she can even eat them!

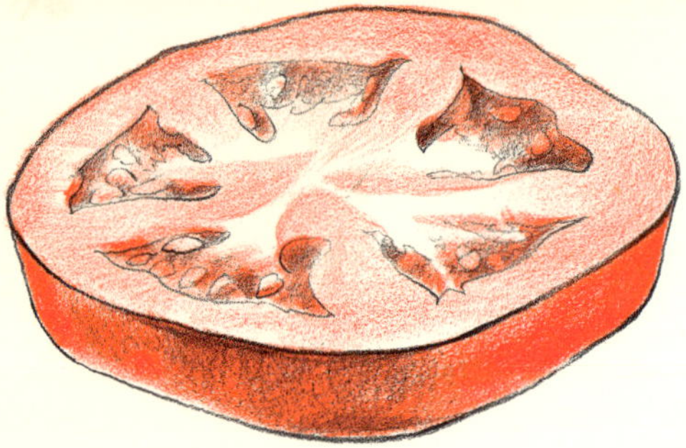

Everyone knows what a slice of tomato looks like. You can see all the seeds very plainly. Sometime, just for fun, count the seeds in one slice. How many would you guess are in the whole tomato? When you eat a tomato, you eat seeds, pod, and everything. When the plant starts growing, it stands up straight, but as the branches grow longer and heavier they bend down and spread out on the ground. All the leaves on the vine shade the dirt underneath, keeping it damp. This helps the plant grow. But if it is very rainy, a lot of tomatoes will rot before they are ripe enough to pick. That is why in some gardens you will see the tomato vines tied up to three or four wooden stakes. This prevents the tomatoes from spoiling on the ground and also makes it much easier to pick them.

Where we live, the cold fall weather comes early, and often there are frosty mornings before all the tomatoes have got red and ripe. Then Daddy pulls up the vine—roots, leaves, tomatoes and all—and puts it in the barn on some hay. The tomatoes keep on ripening inside the barn, where the frost can't turn them black. Isn't that lucky?

One of the best things in my garden isn't a vegetable at all. It isn't the toad. It's a big, happy flower. The seeds from this flower are favorites of all the birds that come to the bird feeder during the winter. It has hundreds of seeds in the center. And there's something else funny about this flower: it turns around during the day! In the morning it is facing the east, and as the sun moves across the sky during the day, the flower keeps turning to face the sun. Now do you know what it is? A sunflower.

A garden is fun even in the winter. Before it snows I spade the garden again. I bury all the corn stalks and broccoli stumps, to rot and make food for next year's plants. But do you know what I leave? I leave the sunflower stalks standing, with their heads bowed down by their heavy loads of seeds. When it snows and the birds can't find anything to eat in all the countryside, they can come to my garden and fill up on sunflower seeds.

Pretty soon winter has come to stay. We feast on all the vegetables stored in neat little boxes in the freezer.

One day the mailman brings the new seed catalogue. With the fine snow blowing outside, and the icicles by the kitchen window getting longer and longer, it's fun to sit in front of the fireplace and look at the beautiful pictures of all the new things you can grow next spring.

And where is toad? He has found a comfortable, deep hole, way down in the ground where it's warm. And there he will sleep, dreaming of next year's garden and wondering if he could manage to eat one of those lovely green tomato worms.